THIS BOOK BELONGS TO:

. . . .

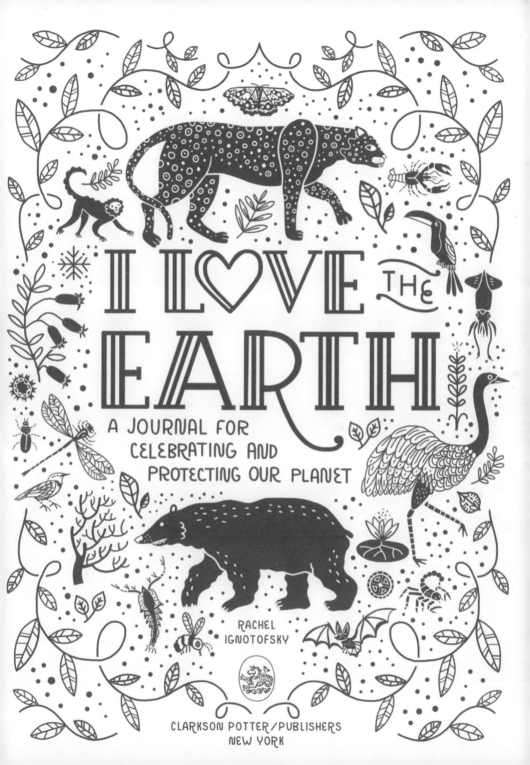

I L♡VE ᴛʜᴇ EARTH

A JOURNAL FOR CELEBRATING AND PROTECTING OUR PLANET

RACHEL IGNOTOFSKY

CLARKSON POTTER/PUBLISHERS
NEW YORK

THE BIG WORLD
WE LIVE IN IS

SMALLER THAN
YOU THINK...

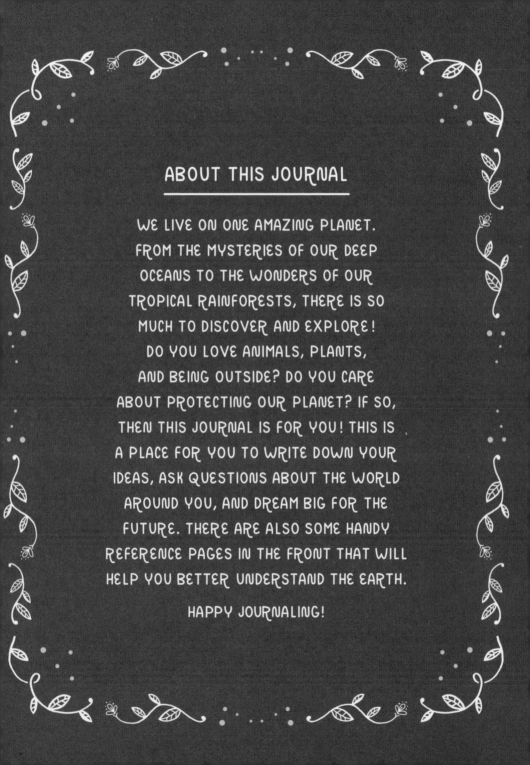

ABOUT THIS JOURNAL

WE LIVE ON ONE AMAZING PLANET.
FROM THE MYSTERIES OF OUR DEEP
OCEANS TO THE WONDERS OF OUR
TROPICAL RAINFORESTS, THERE IS SO
MUCH TO DISCOVER AND EXPLORE!
DO YOU LOVE ANIMALS, PLANTS,
AND BEING OUTSIDE? DO YOU CARE
ABOUT PROTECTING OUR PLANET? IF SO,
THEN THIS JOURNAL IS FOR YOU! THIS IS
A PLACE FOR YOU TO WRITE DOWN YOUR
IDEAS, ASK QUESTIONS ABOUT THE WORLD
AROUND YOU, AND DREAM BIG FOR THE
FUTURE. THERE ARE ALSO SOME HANDY
REFERENCE PAGES IN THE FRONT THAT WILL
HELP YOU BETTER UNDERSTAND THE EARTH.

HAPPY JOURNALING!

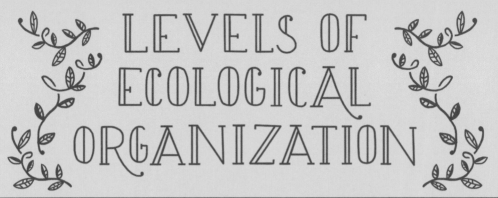

LEVELS OF ECOLOGICAL ORGANIZATION

It is a great big complicated world out there! You can study the entire planet as a whole or study the habits of just one single organism. The *levels of ecology* put it all into context. The largest level is the biosphere, which includes everywhere life is found on Earth. With every level of ecology down from the biosphere, we can zoom in and look at sequentially smaller and more specific parts of the world. The smallest level of ecology is when we study an individual living thing — for example, a squirrel. The levels are like Russian nesting dolls, with each of the six levels fitting inside the next largest level.

BIOSPHERE

Everywhere life
on Earth is found.

BIOME

A region defined by a specific climate
(its temperature and precipitation)
and the animals and plants that have
adapted to survive and thrive in that
type of climate.

ECOSYSTEM

The interactions among all living organisms and their nonliving environment in a certain place.

COMMUNITY

All of the living beings within an ecosystem, such as plants, fungi, animals, and bacteria. Does not include the air, dirt, water, or other nonliving things.

POPULATION

SQUAD GOALS: FIND ACORNS.

A group of individuals of the same species that live within the same community.

INDIVIDUAL

WHERE I LIVE IS MY HABITAT AND HOW I BEHAVE IS MY NICHE.

One specific living organism.

BIOME MAP

Biomes are simply a way to classify and describe general parts of the planet. Each biome is determined by its temperature and precipitation, along with the living things that have evolved in that climate. There are two main types of biomes: terrestrial and aquatic. Ecologists have further broken down those two types into more specific classifications. Biome maps can be divided into many different ways and allow us to understand the similarities between places on opposite sides of the world.

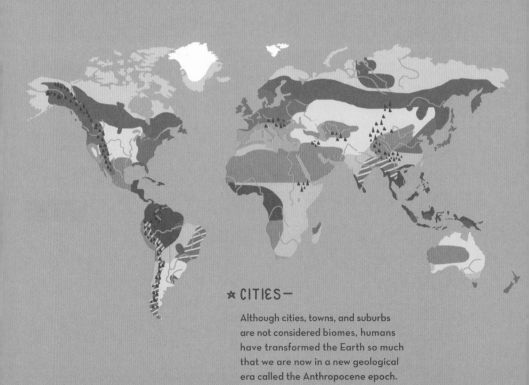

★ CITIES—

Although cities, towns, and suburbs are not considered biomes, humans have transformed the Earth so much that we are now in a new geological era called the Anthropocene epoch.

OCEAN □

MOUNTAIN ▲

GRASSLAND □

FRESH WATER

TAIGA

SCRUBLAND

WETLANDS

TEMPERATE FOREST

SAVANNA

ICE □

TROPICAL RAINFOREST

DESERT

TUNDRA

TROPICAL SEASONAL FOREST

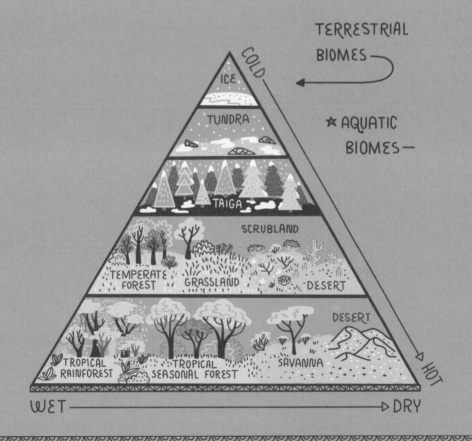

TERRESTRIAL BIOMES

★ AQUATIC BIOMES—

COLD

ICE

TUNDRA

TAIGA

SCRUBLAND

TEMPERATE FOREST

GRASSLAND

DESERT

DESERT

TROPICAL RAINFOREST

TROPICAL SEASONAL FOREST

SAVANNA

HOT

WET → DRY

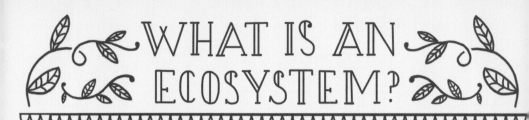

WHAT IS AN ECOSYSTEM?

Every organism on the planet is dependent on others to live. Through ecology, the study of ecosystems, we can begin to grasp how much we rely on the natural world. Ecosystems can range in many sizes, from a large forest to a tiny puddle, and by learning about them, we begin to understand how living organisms in a certain place interact with one another. We also can see how these living things interact with the nonliving parts of their environment (like the soil, the temperature, the air, and the water).

Interactions between wildlife and their environment provide us with important natural services, such as breathable air, fresh water, protection from natural disasters, fertile soil, and, of course, food! By understanding large and small ecosystems, we can see how energy from the sun flows through the food web, and how the cycle of life, death, and decay allows nutrients to be reused. Only when our ecosystems are intact can the natural world continue to seamlessly do the hard work of sustaining life on planet Earth.

FOOD WEB —

The mapping of the flow of energy. Who eats what and who gets energy from whom. Arrows point to who is enjoying a tasty meal, which is the direction energy is moving.

TROPHIC LEVELS —

An organism's position in the food web, and how far away it is from the original source of energy (the sun), starting with producers and typically ending with apex predators.

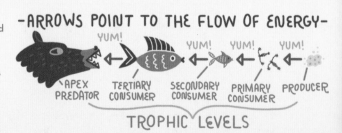

-ARROWS POINT TO THE FLOW OF ENERGY-

YUM!   YUM!   YUM!   YUM!

APEX PREDATOR   TERTIARY CONSUMER   SECONDARY CONSUMER   PRIMARY CONSUMER   PRODUCER

TROPHIC LEVELS

WHO EATS WHAT —

Producers make their own food from solar energy. Herbivores eat only plants. Carnivores eat only other animals. Omnivores eat both plants and animals. Decomposers eat waste and dead organisms.

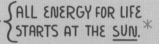

ALL ENERGY FOR LIFE STARTS AT THE SUN.*

*SOME MICROSCOPIC LIFE GETS ITS ENERGY FROM THERMAL VENTS.

PRODUCER

PRIMARY CONSUMER

APEX PREDATORS

WEATHER

CLIMATE

PRODUCERS (PLANTS)

ABIOTIC (NOT LIVING THINGS)

ROCKS

TERTIARY CONSUMER

WATER

SECONDARY CONSUMER

SOIL

DECOMPOSERS

PRIMARY CONSUMER

THE CLASSIFICATION OF LIVING THINGS

Taxonomic ranking helps scientists classify and identify different species. Scientists include every single living thing that has ever existed on Earth, which allows us to see how life on Earth has evolved. Taxonomic ranking also helps us understand what different species have in common—even if they have been extinct for thousands of years or live on opposite sides of the world!

— THE MAIN DOMAINS —

BACTERIA

SINGLE CELL ORGANISMS WITH NO DEFINED NUCLEUS

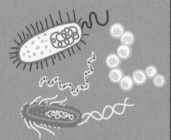

ARCHAEA

SINGLE CELL ORGANISMS WITH NO DEFINED NUCLEUS & DIFFERENT BIOCHEMISTRY THAN BACTERIA.

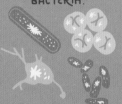

— EUKARYA —

ORGANISMS THAT HAVE CELLS WITH A NUCLEUS

ANIMALIA

PLANTAE

FUNGI

PROTISTA

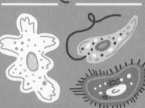

— LEVELS OF CLASSIFICATION —

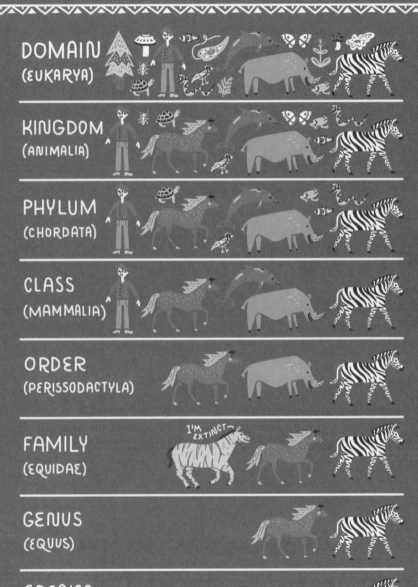

DOMAIN
(EUKARYA)

KINGDOM
(ANIMALIA)

PHYLUM
(CHORDATA)

CLASS
(MAMMALIA)

ORDER
(PERISSODACTYLA)

FAMILY
(EQUIDAE)

I'M EXTINCT

GENUS
(EQUUS)

SPECIES
(ZEBRA)

HOW LIVING THINGS INTERACT

Competing for food and resources, finding a place to call home, and reproducing are some of the main priorities for all species. To do this, animals, bacteria, and plants have evolved to interact in many different ways to survive. These interactions help to maintain a balanced and healthy ecosystem.

YIKES!

PREDATION: ONE SPECIES EATS THE OTHER.

COMMENSALISM: ONE SPECIES GAINS AND THE OTHER IS UNAFFECTED.

FREE RIDE!

HOW TO PROTECT OUR PLANET

There are many things we can do to preserve the natural world. Never forget that you have the power to protect our planet!

EDUCATE

We need to understand how our ecosystems work in order to protect them.

>>>> RECYCLE AND REUSE <<<<

Don't just throw away broken things.
Repair them or turn them into something new!

PLANT TREES

Trees and forests filter greenhouse gasses and create oxygen.

CONSERVE -WATER-

TURN OFF WATER

Fresh water is a limited resource, and it's scarce in many parts of the world. Using less water also leads to less runoff and waste water dumped into the ocean.

✕✕✕✕ HOW WE DO BUSINESS ✕✕✕✕

Often, clothes, electronics, and other products are created to be used, thrown away, and replaced. This is a waste of valuable resources. Instead, demand that companies create products that are made to last a long time and can be repaired. Buy from businesses that put the planet first!

▶▶▶▶▶ VOLUNTEER! ◀◀◀◀◀

Conservation groups need your help.

COMPOST PAPER PLASTIC GLASS METAL

····ZERO LANDFILL WASTE····

Recycling in your home is great, but to have a larger impact it needs to happen on a bigger scale. Help create systems for everyone to compost and recycle at your work or school.

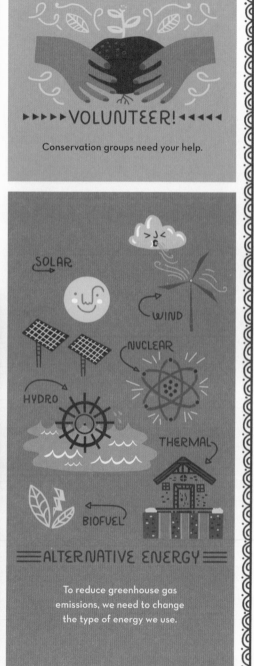

SOLAR

WIND

NUCLEAR

HYDRO

THERMAL

BIOFUEL

═ALTERNATIVE ENERGY═

To reduce greenhouse gas emissions, we need to change the type of energy we use.

HOW TO PROTECT OUR PLANET

SUSTAINABLE FISHING

Our entire world depends on the marine ecosystems. We need to end overfishing and only fish responsibly.

▲▽▲ REGULATIONS ▲▽▲

We need to create and enforce regulations that prevent farms and factories from polluting our streams, oceans, and air.

▸▸▸ EAT LESS MEAT ◂◂◂

It takes more energy and resources to raise livestock than to grow crops. Reducing your meat and fish consumption helps the whole world.

SUSTAINABLE FARMING

The huge and growing human population will always need large-scale farming, but with a knowledge of ecology, biology, and economics, we can invest in making large-scale agriculture profitable and healthy for the whole world!

SUSTAINABLE WORK $

CLEAN WATER

— FOOD — SECURITY

FIGHT POVERTY

When people in poverty have few options, they can turn to illegal poaching, lumber exploitation, unsustainable farming and herding, and dangerous mining. We cannot expect poor people to shoulder the responsibility of saving the planet when they are worried about providing for themselves. By addressing the underlying problems of poverty, we can all find a way to live, survive, and thrive without harming our planet.

PROTECTED OCEAN

NATIONAL PARK

NATURE PRESERVE

PROTECT WILDLIFE

To preserve important ecosystems, we need to protect natural wildlife areas.

REDUCE YOUR CARBON FOOTPRINT

Use less fossil fuel and coal!
Use less electricity!
Carpool and drive less!
Ride your bike!
Use less plastic!

CALL YOUR REPRESENTATIVES

VOTE

SPEAK UP!

RAISE YOUR VOICE

Get out there and demand the change you want to see in the world.

SHARE YOUR KNOWLEDGE!

"LOOK DEEP INTO NATURE AND YOU WILL
UNDERSTAND EVERYTHING BETTER."
—ALBERT EINSTEIN, NOBEL PRIZE-WINNING PHYSICIST

HOW DOES BEING IN NATURE MAKE YOU FEEL?
WHAT IS YOUR FAVORITE PART ABOUT BEING OUTSIDE?

"ONLY IF WE UNDERSTAND, WILL WE CARE. ONLY IF WE CARE, WILL WE HELP. ONLY IF WE HELP SHALL ALL BE SAVED."
—JANE GOODALL, PRIMATOLOGIST, ETHNOLOGIST, AND ANTHROPOLOGIST

_____

_____

_____

_____

_____

_____

_____

_____

_____

_____

_____

_____

_____

_____

_____

IF IT WERE UP TO YOU, HOW WOULD YOU SOLVE
SOME OF THE WORLD'S BIGGEST PROBLEMS?

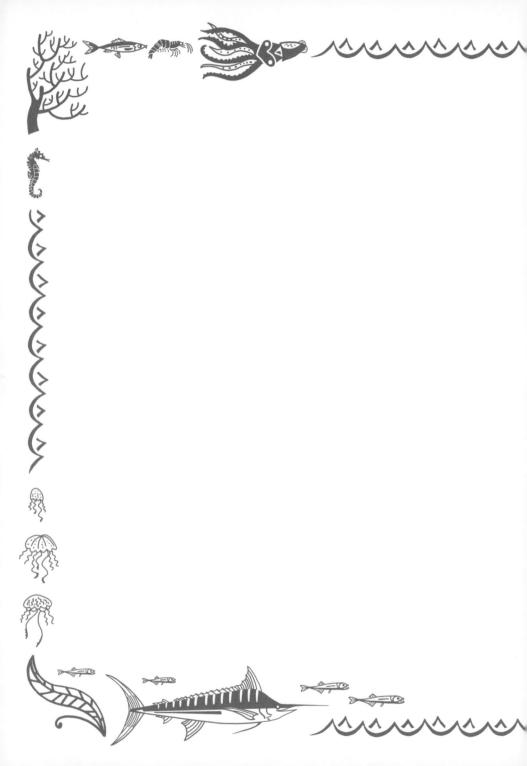

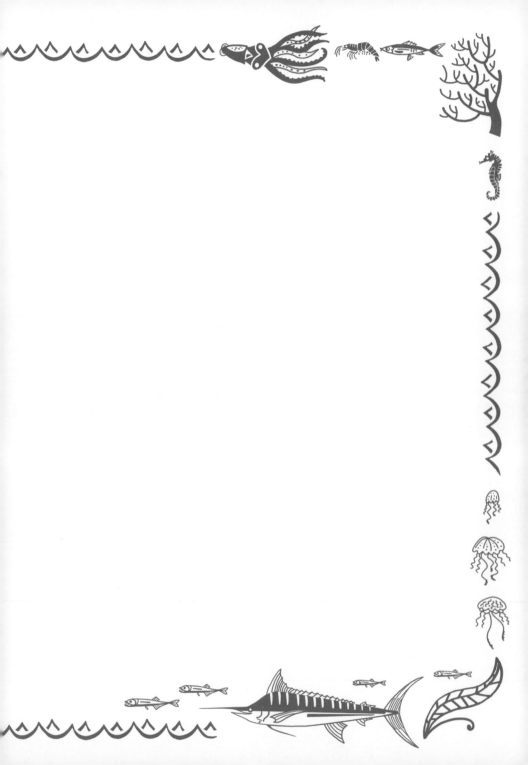

"NATURE IS NOT A PLACE TO VISIT. IT IS HOME."
—GARY SNYDER, POET AND ENVIRONMENTALIST

WHAT ARE SOME OF YOUR FAVORITE WILD ANIMALS?
HAVE YOU EVER SEEN THEM IN REAL LIFE?

"WE NEED TO RESPECT THE OCEANS AND TAKE CARE OF THEM AS IF OUR LIVES DEPENDED ON IT. BECAUSE THEY DO."
—SYLVIA EARLE, MARINE BIOLOGIST, EXPLORER, AND AQUANAUT

WHAT ARE YOU REALLY GOOD AT? WHAT ARE
SOME THINGS YOU WANT TO GET BETTER AT?

"NATURE DOES NOT HURRY, YET EVERYTHING
IS ACCOMPLISHED."

—LAO TZU, ANCIENT PHILOSOPHER

LIST ALL THE DIFFERENT PLACES YOU HAVE TRAVELED TO.
WHAT WAS DIFFERENT ABOUT THE PLANTS AND ANIMALS
IN EACH OF THESE PLACES?

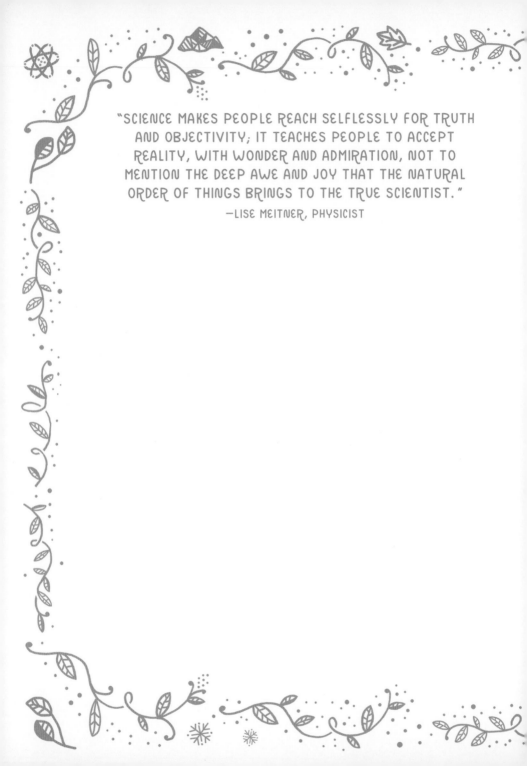

"SCIENCE MAKES PEOPLE REACH SELFLESSLY FOR TRUTH AND OBJECTIVITY; IT TEACHES PEOPLE TO ACCEPT REALITY, WITH WONDER AND ADMIRATION, NOT TO MENTION THE DEEP AWE AND JOY THAT THE NATURAL ORDER OF THINGS BRINGS TO THE TRUE SCIENTIST."

—LISE MEITNER, PHYSICIST

WHAT KIND OF ANIMALS, PLANTS, AND NATURAL FEATURES
HAVE YOU OBSERVED IN YOUR NEIGHBORHOOD?

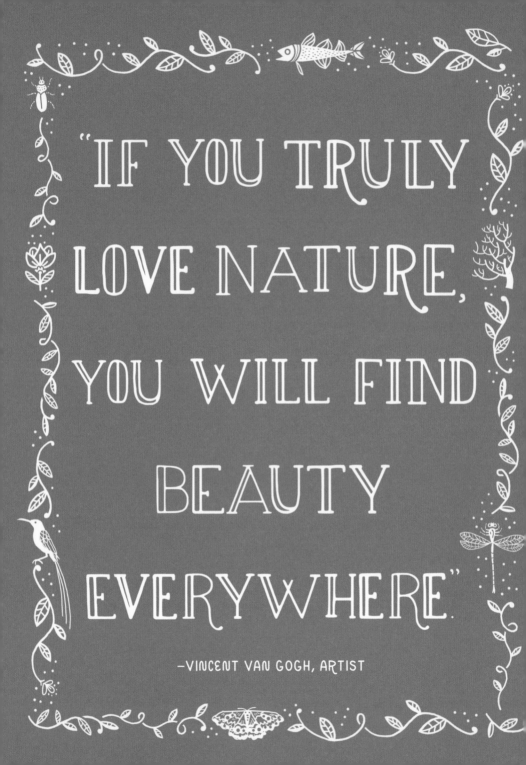

"IF YOU TRULY LOVE NATURE, YOU WILL FIND BEAUTY EVERYWHERE."

—VINCENT VAN GOGH, ARTIST

WHAT ARE SOME THINGS PEOPLE CAN DO TO HELP
PROTECT THE EARTH? DO YOU DO THEM? DO PEOPLE
IN YOUR FAMILY, SCHOOL, AND COMMUNITY DO THEM?

_____

_____

_____

_____

_____

_____

_____

_____

_____

_____

_____

_____

_____

_____

_____

_____

HAVE YOU EVER HELPED TAKE CARE OF A PLANT
OR WORKED IN A GARDEN? WHAT WAS IT LIKE?

"IN THE SPRING, AT THE END OF THE DAY,
YOU SHOULD SMELL LIKE DIRT."
—MARGARET ATWOOD, AUTHOR, POET, AND ENVIRONMENTAL ACTIVIST

HAVE YOU EVER SWAM IN A NATURAL BODY OF WATER
LIKE AN OCEAN, A LAKE, OR A RIVER? WHAT WAS IT LIKE?
WOULD YOU WANT TO DO IT AGAIN?

USE THIS PAGE TO DRAW OBSERVATIONS FROM NATURE.
WHAT DO NATURAL OBJECTS LOOK LIKE CLOSE-UP? DRAW THE
TEXTURE OF ROCKS, LEAVES, FEATHERS, OR SEASHELLS.

"NATURE WILL BEAR THE CLOSEST INSPECTION. SHE INVITES
US TO LAY OUR EYE LEVEL WITH HER SMALLEST LEAF,
AND TAKE AN INSECT VIEW OF ITS PLAIN."
—HENRY DAVID THOREAU, NATURALIST, PHILOSOPHER, AND AUTHOR

WHO ARE YOUR ROLE MODELS AND WHY?

_____

_____

_____

_____

_____

_____

_____

_____

_____

_____

_____

_____

_____

_____

_____

_____

_____

_____

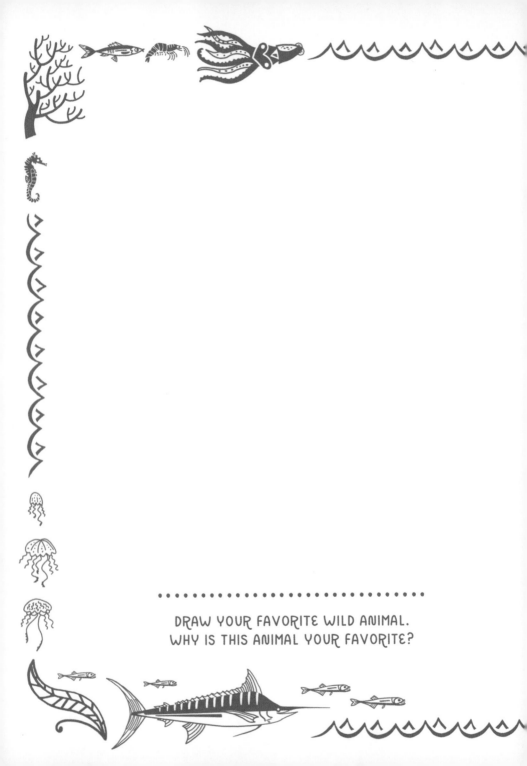

DRAW YOUR FAVORITE WILD ANIMAL.
WHY IS THIS ANIMAL YOUR FAVORITE?

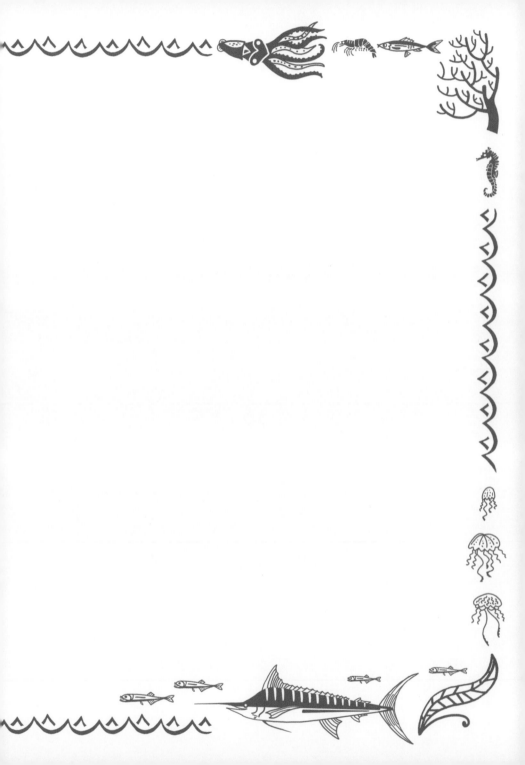

"ALL MY LIFE THROUGH, THE NEW SIGHTS OF
NATURE MADE ME REJOICE LIKE A CHILD."

—MARIE CURIE, NOBEL PRIZE-WINNING PHYSICIST AND CHEMIST

HAVE YOU EVER HAD TO STAND UP FOR WHAT
YOU BELIEVE IN? WHAT WAS THAT LIKE?

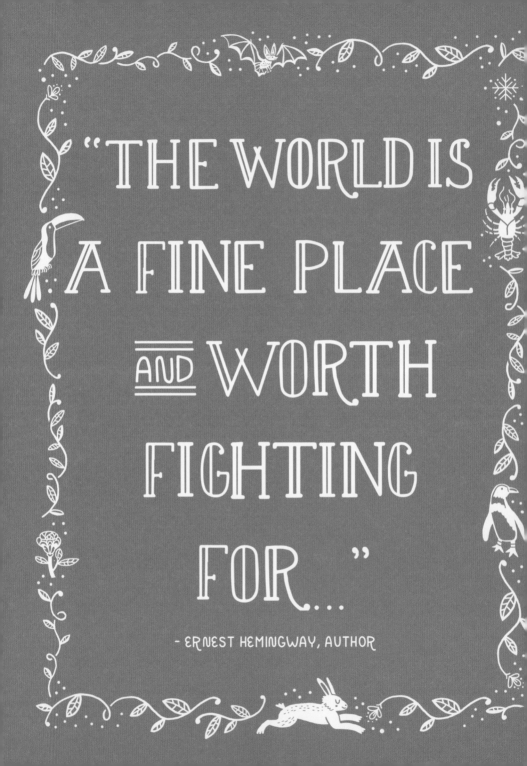

"THE WORLD IS A FINE PLACE AND WORTH FIGHTING FOR..."

- ERNEST HEMINGWAY, AUTHOR

MAKE A LIST OF ALL THE PLACES IN THE WORLD
THAT YOU WANT TO VISIT SOMEDAY.
WHAT INTERESTS YOU ABOUT THESE PLACES?

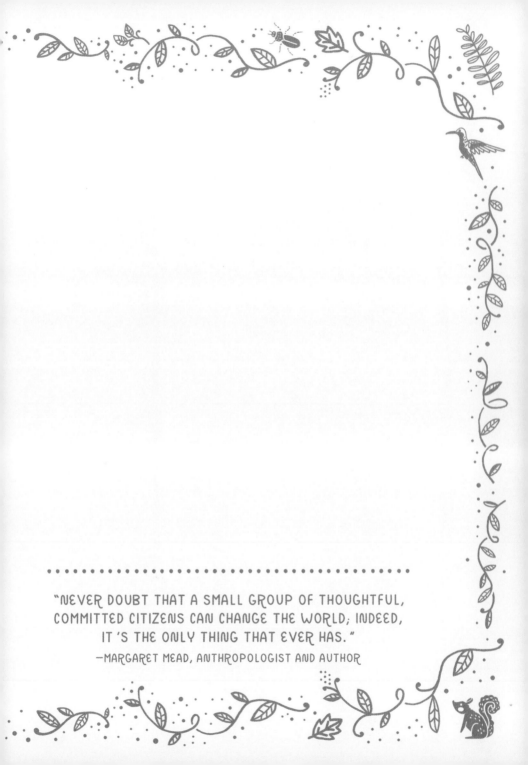

"NEVER DOUBT THAT A SMALL GROUP OF THOUGHTFUL,
COMMITTED CITIZENS CAN CHANGE THE WORLD; INDEED,
IT'S THE ONLY THING THAT EVER HAS."
—MARGARET MEAD, ANTHROPOLOGIST AND AUTHOR

HOW DO YOU REDUCE, REUSE, AND RECYCLE
IN YOUR DAILY LIFE? HOW COULD YOU DO MORE?

WHAT DO YOU THINK MAKES A HAPPY LIFE?

"AN UNDERSTANDING OF THE NATURAL WORLD AND WHAT'S IN IT IS A SOURCE OF NOT ONLY A GREAT CURIOSITY BUT GREAT FULFILLMENT."
—DAVID ATTENBOROUGH, NATURAL HISTORIAN

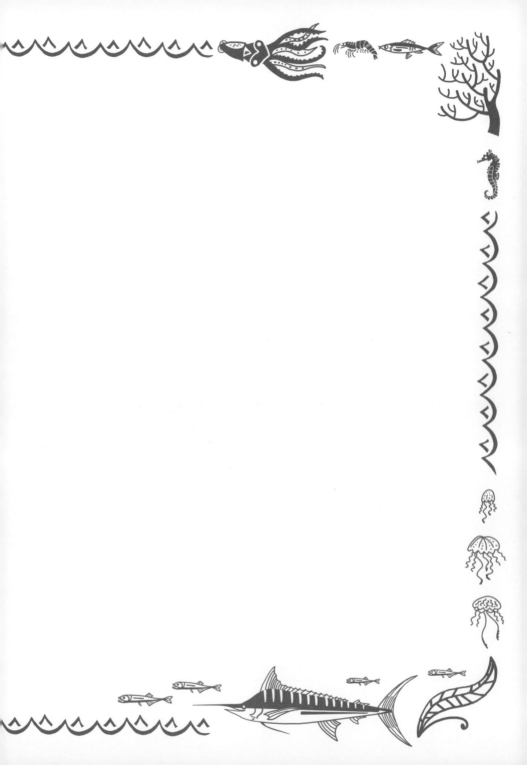

ANIMALS HAVE DEVELOPED SPECIAL ABILITIES TO SURVIVE
IN DIFFERENT ENVIRONMENTS. SOME ANIMALS HAVE
EXTREME NIGHT VISION, OTHER ANIMALS CAN MOVE AS FAST
AS A SPORTS CAR. WHAT ANIMAL SUPER POWER
DO YOU WISH YOU HAD AND WHY?

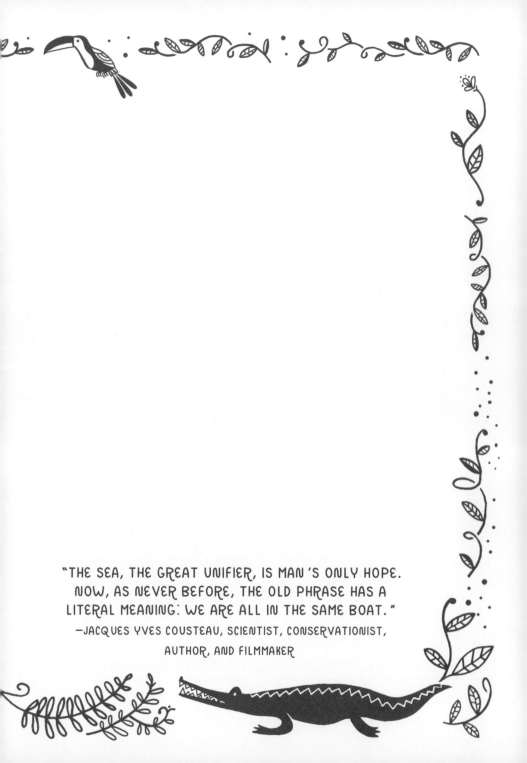

"THE SEA, THE GREAT UNIFIER, IS MAN'S ONLY HOPE.
NOW, AS NEVER BEFORE, THE OLD PHRASE HAS A
LITERAL MEANING: WE ARE ALL IN THE SAME BOAT."
—JACQUES YVES COUSTEAU, SCIENTIST, CONSERVATIONIST,
AUTHOR, AND FILMMAKER

OBSERVE FROM NATURE: HOW DOES THE WILDLIFE
IN YOUR BACKYARD OR NEIGHBORHOOD CHANGE FROM
DAYTIME TO NIGHTTIME. DRAW WHAT YOU SEE.

PRESS A FEW LEAVES AND/OR SMALL
FLOWERS BETWEEN THESE PAGES.

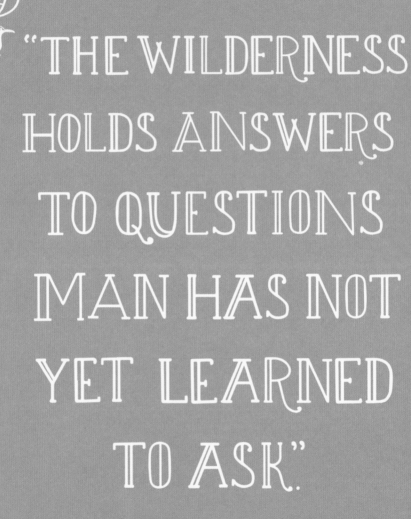

"THE WILDERNESS HOLDS ANSWERS TO QUESTIONS MAN HAS NOT YET LEARNED TO ASK."

— NANCY NEWHALL, EDITOR AND PHOTOGRAPHY CRITIC

WHAT ARE SOME OF THE WAYS HUMANS HAVE HURT
THE NATURAL WORLD? HOW DOES IT MAKE YOU FEEL?

"THE MORE CLEARLY WE CAN FOCUS OUR
ATTENTION ON THE WONDERS AND REALITIES
OF THE UNIVERSE ABOUT US, THE LESS TASTE
WE SHALL HAVE FOR DESTRUCTION."
—RACHEL CARSON, MARINE BIOLOGIST,
CONSERVATIONIST, AND AUTHOR

COLOR THIS PAGE!

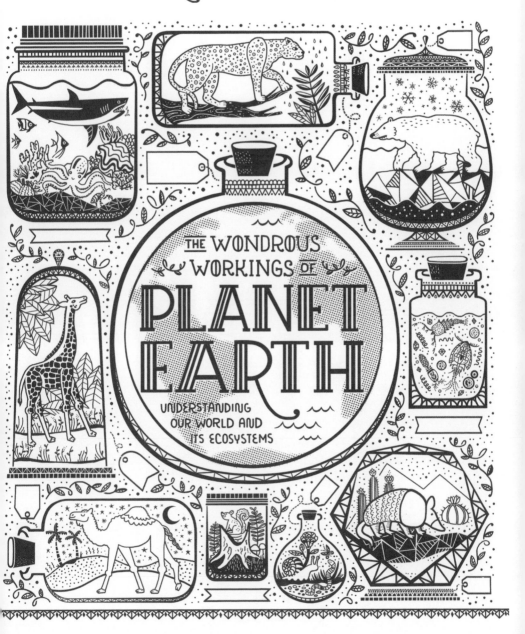

THE WONDROUS WORKINGS OF **PLANET EARTH**

UNDERSTANDING OUR WORLD AND ITS ECOSYSTEMS

"STUDY NATURE, LOVE NATURE, STAY CLOSE
TO NATURE. IT WILL NEVER FAIL YOU."
—FRANK LLOYD WRIGHT, ARCHITECT, DESIGNER, AND EDUCATOR

IF YOU COULD EXPLORE ANYWHERE ON THE PLANET,
WHERE WOULD YOU GO? WHAT DO YOU THINK
YOU WOULD DISCOVER THERE?

"EACH SPECIES IS A MASTERPIECE, A CREATION ASSEMBLED
WITH EXTREME CARE AND GENIUS."

—E. O. WILSON, BIOLOGIST, NATURALIST, AND AUTHOR

WHAT ARE SOME OF YOUR DREAMS FOR THE FUTURE?
HOW CAN YOU MAKE THOSE DREAMS A REALITY?

WOULD YOU RATHER CLIMB MOUNTAINS OR SCUBA DIVE
IN THE OCEAN? WHAT WOULD YOU WANT TO SEE AND DO
ON TOP OF A MOUNTAIN OR UNDER THE SEA?

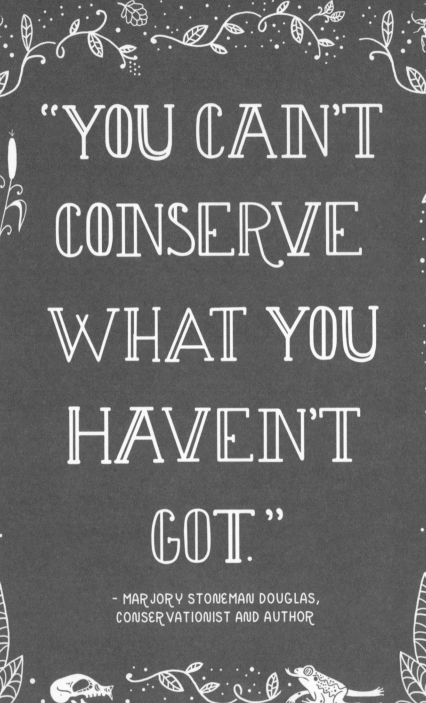

"YOU CAN'T CONSERVE WHAT YOU HAVEN'T GOT."

- MARJORY STONEMAN DOUGLAS, CONSERVATIONIST AND AUTHOR

GO OUTSIDE AND SEE IF YOU CAN SPOT ANY BIRDS.
WHAT KINDS DO YOU SEE? WHAT ARE THEY DOING?
ARE THEY MAKING ANY SOUNDS?
DESCRIBE AND DRAW WHAT YOU SEE.

"EARTH PROVIDES ENOUGH TO SATISFY EVERY MAN'S
NEED, BUT NOT EVERY MAN'S GREED."
—MAHATMA GANDHI, ACTIVIST AND INDEPENDENCE LEADER

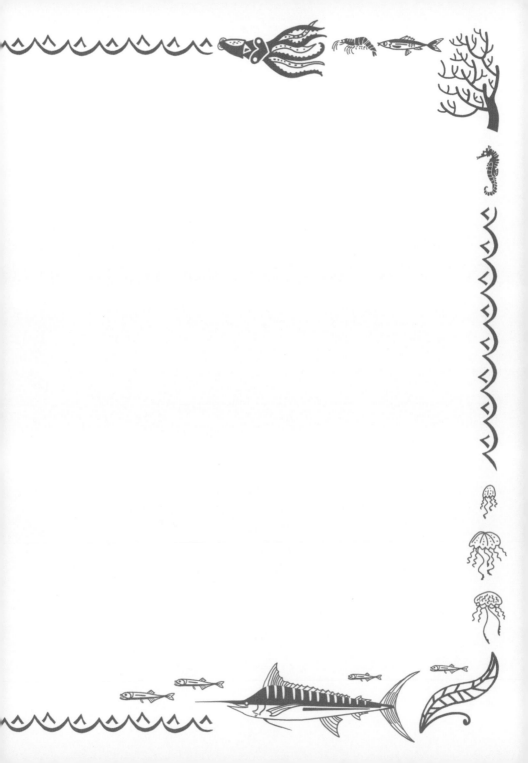

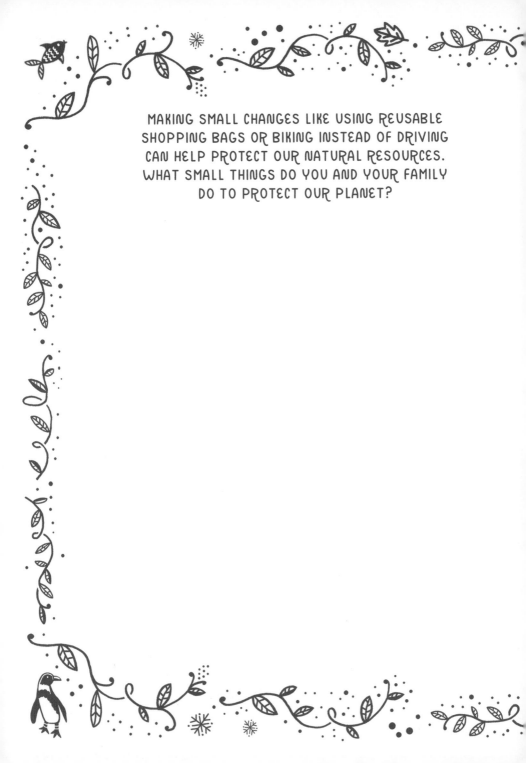

MAKING SMALL CHANGES LIKE USING REUSABLE SHOPPING BAGS OR BIKING INSTEAD OF DRIVING CAN HELP PROTECT OUR NATURAL RESOURCES. WHAT SMALL THINGS DO YOU AND YOUR FAMILY DO TO PROTECT OUR PLANET?

"THE EARTH IS WHAT WE ALL HAVE IN COMMON."
—WENDELL BERRY, POET AND ENVIRONMENTALIST

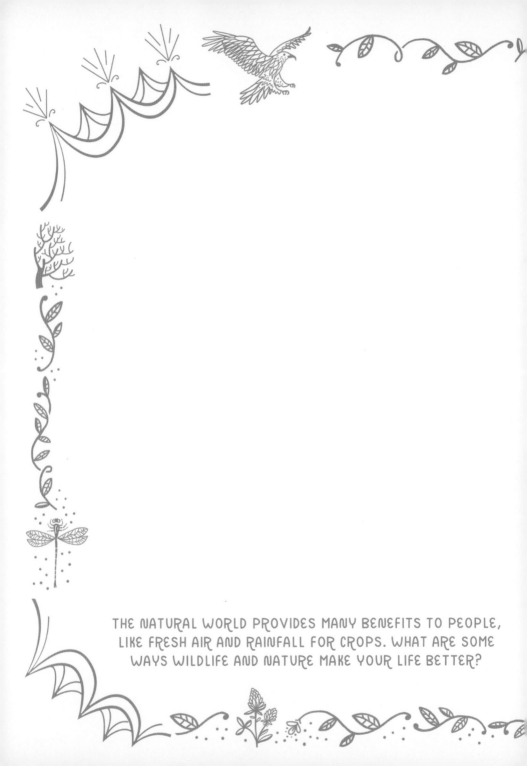

THE NATURAL WORLD PROVIDES MANY BENEFITS TO PEOPLE,
LIKE FRESH AIR AND RAINFALL FOR CROPS. WHAT ARE SOME
WAYS WILDLIFE AND NATURE MAKE YOUR LIFE BETTER?

"ADOPT THE PACE OF NATURE: HER SECRET IS PATIENCE."
—RALPH WALDO EMERSON, PHILOSOPHER, POET, AND AUTHOR

COLOR THIS PAGE!

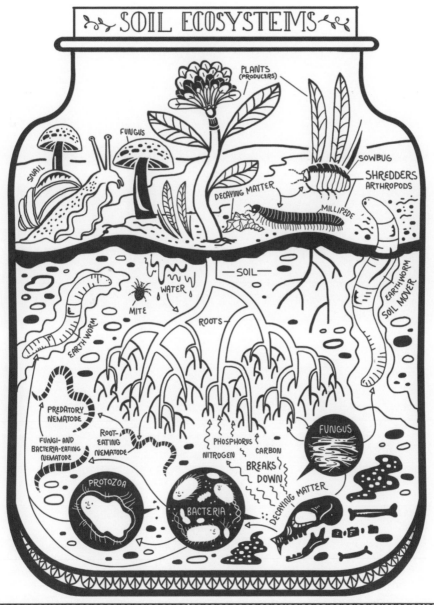

SOIL ECOSYSTEMS

"NOTHING IS MORE BEAUTIFUL THAN THE LOVELINESS OF THE WOODS BEFORE SUNRISE."
—GEORGE WASHINGTON CARVER, AGRICULTURAL SCIENTIST AND INVENTOR

DO YOU HAVE ANY PETS? WHAT HAVE YOU LEARNED
ABOUT ANIMALS FROM LIVING WITH THEM?

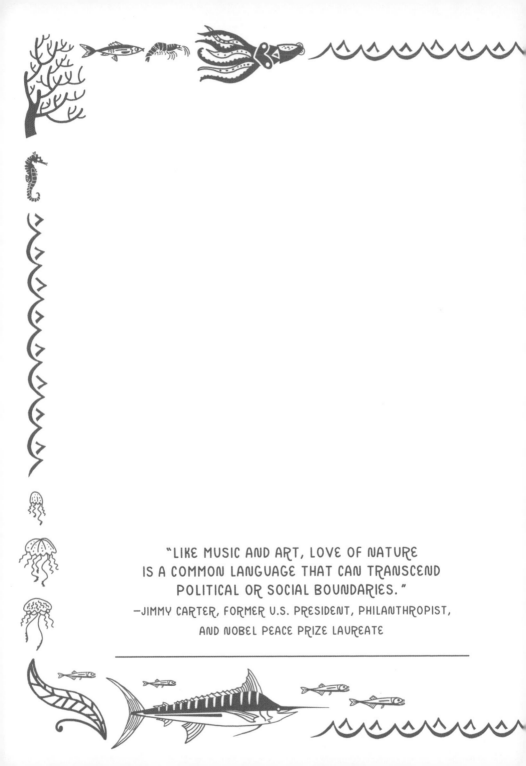

"LIKE MUSIC AND ART, LOVE OF NATURE
IS A COMMON LANGUAGE THAT CAN TRANSCEND
POLITICAL OR SOCIAL BOUNDARIES."
—JIMMY CARTER, FORMER U.S. PRESIDENT, PHILANTHROPIST,
AND NOBEL PEACE PRIZE LAUREATE

"GRASS MADE IT POSSIBLE FOR THE HUMAN RACE
TO ABANDON CAVE LIFE AND FOLLOW HERDS."
—MARY AGNES CHASE, BOTANIST AND SUFFRAGIST

WHAT ARE SOME ACTIVITIES YOU LIKE TO DO OUTSIDE?

"THE WILDERNESS AND THE IDEA OF WILDERNESS IS ONE OF THE PERMANENT HOMES OF THE HUMAN SPIRIT."

—JOSEPH WOOD KRUTCH, WRITER, CRITIC, AND NATURALIST

THE WORD HABITAT MEANS THE NATURAL HOME
OF A LIVING ORGANISM. WHAT'S YOUR HABITAT?

DO YOU HAVE ANY FEARS THAT YOU'VE OVERCOME?
HOW DID YOU DO IT? ARE THERE OTHER FEARS
YOU WANT TO OVERCOME?

_____

"IT IS AN INCALCULABLE ADDED PLEASURE TO ANY ONE'S SUM OF HAPPINESS IF HE OR SHE GROWS TO KNOW, EVEN SLIGHTLY AND IMPERFECTLY, HOW TO READ AND ENJOY THE WONDER-BOOK OF NATURE."

—THEODORE ROOSEVELT,
U.S. PRESIDENT AND CONSERVATIONIST

COLOR THIS PAGE!

GREAT HORNED OWL

APEX PREDATOR MOUNTAIN LION

COASTAL REDWOOD TREE

PRODUCER COASTAL REDWOOD TREE

OMNIVORE AMERICAN BLACK BEAR

DOUGLAS FIR TREE

YELLOW-CHEEKED CHIPMUNK

SALMON

PRIMARY CONSUMER BLACK-TAILED DEER

ACORN

ROOSEVELT ELK

LICHENS

FALLEN GIANT REDWOOD TREE

TURKEY TAIL MUSHROOM

STRIPED SKUNK

LICHEN

ALGAL CELL

INSECTS

GRASSES

FUNGAL HYPHAE

DECOMPOSERS

WHAT ADVICE WOULD YOU GIVE TO A
FRIEND WHO IS HAVING A HARD TIME?

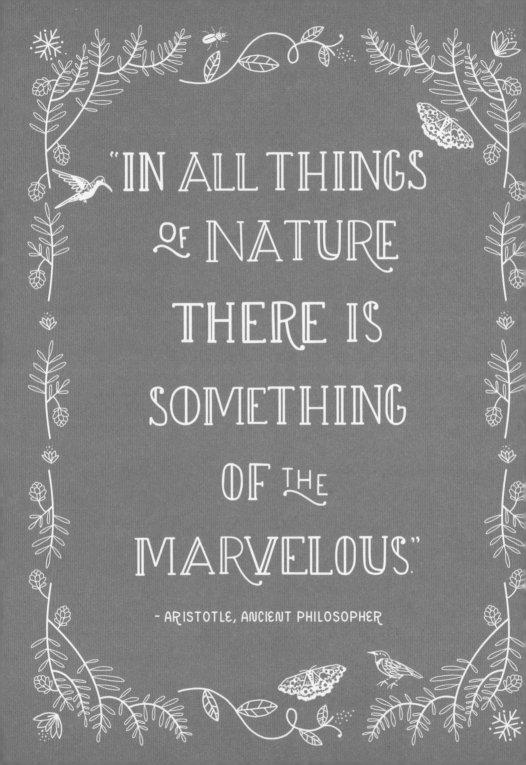

"IN ALL THINGS OF NATURE THERE IS SOMETHING OF THE MARVELOUS."

– ARISTOTLE, ANCIENT PHILOSOPHER

HOW CAN PEOPLE HELP PROTECT ENDANGERED SPECIES?
IS THERE ANYTHING YOU CAN DO TO HELP?

"WE INTER-BREATHE WITH THE RAIN FORESTS, WE DRINK FROM THE OCEANS. THEY ARE PART OF OUR OWN BODY."

—THICH NHAT HANH, BUDDHIST MONK AND PEACE ACTIVIST

OBSERVE FROM NATURE: HOW DOES THE SKY
CHANGE DURING DIFFERENT TYPES OF WEATHER?
DRAW WHAT YOU SEE.

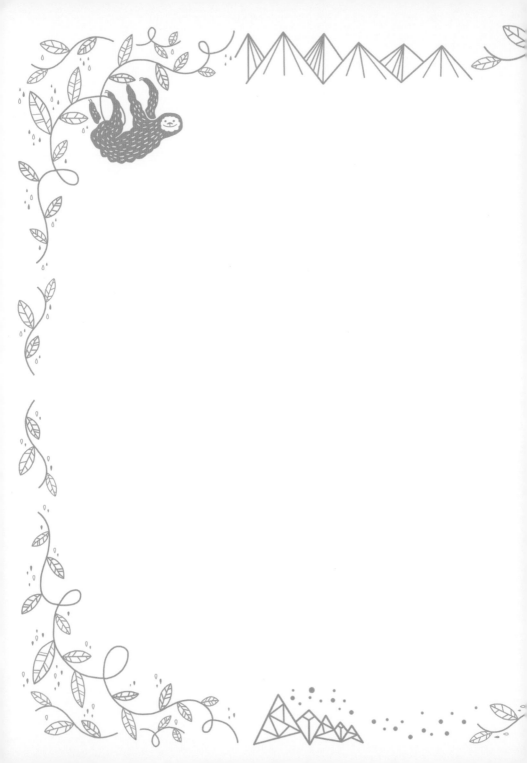

"WHEN ONE TUGS AT A SINGLE THING IN NATURE,
HE FINDS HE IS ATTACHED TO THE REST OF THE WORLD."
—JOHN MUIR, NATURALIST, CONSERVATIONIST, AND AUTHOR

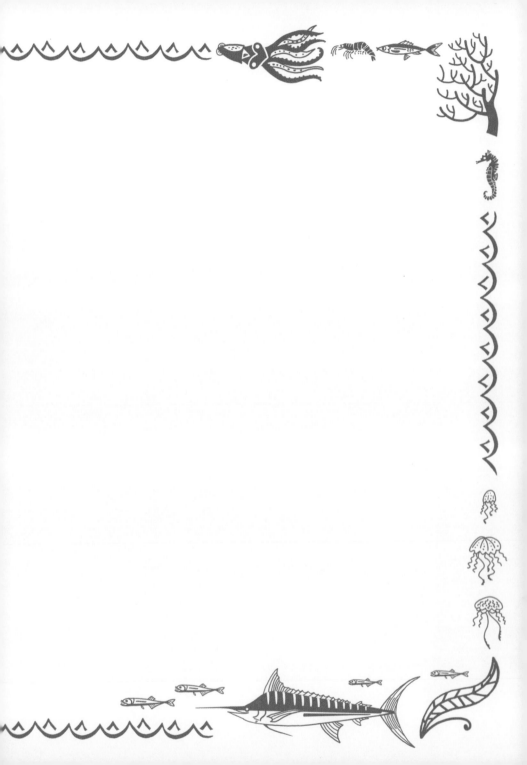

PUBLISHED IN THE UNITED STATES BY CLARKSON POTTER/
PUBLISHERS, AN IMPRINT OF RANDOM HOUSE, A DIVISION
OF PENGUIN RANDOM HOUSE LLC, NEW YORK.

CLARKSONPOTTER.COM

CLARKSON POTTER IS A TRADEMARK AND POTTER WITH COLOPHON
IS A REGISTERED TRADEMARK OF PENGUIN RANDOM HOUSE LLC.

PENGUIN RANDOM HOUSE IS COMMITTED TO A SUSTAINABLE FUTURE
FOR OUR BUSINESS, OUR READERS AND OUR PLANET. THIS BOOK
IS MADE FROM FOREST STEWARDSHIP COUNCIL® CERTIFIED PAPER.

ISBN 978-0-593-13503-7

PRINTED IN CHINA

BOOK DESIGN BY ROOPA SACHIDANAND
ILLUSTRATIONS BY RACHEL IGNOTOFSKY

WRITTEN BY KAITLIN KETCHUM AND RACHEL IGNOTOFSKY

BASED ON THE WONDROUS WORKINGS OF PLANET EARTH (TEN SPEED PRESS, 2018)

10 9 8 7 6 5 4 3 2 1

FIRST EDITION